REDUCE, REUSE, RECYCLE!

Clothes and Toys

Deborah Chancellor

WAYLAND

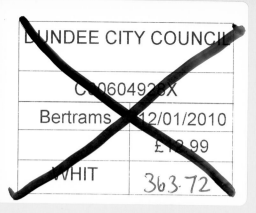

First published in 2009 by Wayland

Copyright © Wayland 2009

Wayland
338 Euston Road
London NW1 3BH

Wayland Australia
Level 17/207 Kent Street
Sydney NSW 2000

Editor: Katie Powell
Designer: Elaine Wilkinson
Consultant: Kate Ruttle
Picture Researcher: Shelley Noronha
Photographer: Andy Crawford

British Library Cataloguing in Publication Data

Chancellor, Deborah
Clothes and toys. - (Reduce, reuse, recycle!)
1. Recycling (Waste, etc.) - Juvenile literature 2. Waste
minimization - Juvenile literature 3. Textile fabrics -
Recycling - Juvenile literature
I. Title
363.7'288

ISBN 978 0 7502 5713 8

Cover: © Ecoscene/Ed Maynard.

pg 1 Wayland, 2 Recycle Now, 4 Patrick Bennett/CORBIS, 5 Brad Killer/Istock, 6 Sean Locke/Istock, 7 Recycle Now,
8 Ed Maynard/ Ecoscene, 8 Recycle Now, 10 Christine Osborne/Ecoscene, 11 James L. Amos/CORBIS, 12 Istock,
13 Recycle Now, 14 Wayland, 15 moodboard/Corbis, 16 Wayland, 17 Peter Morgan/Reuters/Corbis, 18 Corbis Sygma,
19 Wayland, 20 Peter Usbeck /Alamy, 21 Uwe Krejci/zefa/Corbis, 22 Jupiterimages/Ablestock/Alamy, 23 Patrick
Robert/Sygma/Corbis, 24 Getty images, 25 Istock, 26 Recycle Now, 27 Melvyn Longhurst / Alamy, 28, 29 Wayland,
Cover: Ed Maynard/Ecoscene.

With thanks to RecycleNow for kind permission to reproduce the photographs on the imprint page, 7, 9, 13 and 26.

The author and publisher would like to thank the following models: Lawrence Do of Scotts Park Primary School,
Sam Mears and Madhvi Paul.

Printed in China

Wayland is a division of Hachette Children's Books, an Hachette UK company.
www.hachettelivre.co.uk

Contents

What a waste! 4

Reduce, reuse, recycle 6

Materials for clothes 8

Saving clothes 10

Wear it again 12

Recycling clothes 14

Too many toys 16

Toy materials 18

Toy packaging 20

Pass it on 22

Sell, swap or borrow 24

Toy recycling 26

Make a multi-coloured coat 28

Further information 30

Glossary 31

Index 32

Words in **bold** can be found in the glossary.

What a waste!

Every day, we throw away huge amounts of rubbish. Most of it is taken to rubbish dumps called **landfill sites**, where it is crushed and buried. A small amount of our rubbish is burnt in big **furnaces** called **incinerators**.

◀ Rubbish is collected from bins outside our homes in waste collection lorries.

Did You Know?

In the United Kingdom (UK), most children get more than ten toys a year. This adds up to 65 million new toys.

We can **reduce** waste by being careful about what we throw out. A lot of toys and clothes don't need to be thrown away. If they are in a good **condition**, they can be **reused** by someone else. If this isn't possible, maybe they can be **recycled** and turned into something new.

◀ *This woman is making a new bag from some old clothes.*

5

Reduce, reuse, recycle

Clothes and toys do not last forever. People grow out of them, or they get worn out. You can reduce the amount of toys and clothes you throw away. Before you get rid of anything, think how it could be reused or recycled.

◄ *Some clothes that are bought are never worn. This is very wasteful.*

Giving toys and clothes away when they are not wanted anymore means they can be reused by someone else. If they are damaged, it may be possible to take them to a **recycling bank**. They will be collected from there and sorted at a **recycling centre**.

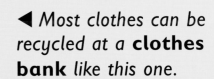

◀ *Most clothes can be recycled at a* **clothes bank** *like this one.*

Did You Know? More than half of the world's **population** wear second-hand clothes. Think before you throw something away. Someone else may need it, somewhere else in the world.

Materials for clothes

Materials that are made from threads are called **textiles**. There are two kinds of textiles – **natural** and **man-made**. Cotton, wool and silk are natural textiles. Man-made textiles include nylon and polyester. Nearly all textiles can be recycled.

◀ *Unwanted clothes are being sorted at this recycling centre.*

You Can Help!

When you take shoes to a shoe bank, tie the laces together or put an elastic band around them, so they don't get separated.

Another natural material is leather. Leather is made from the skin of animals, such as cows. It is often used to make shoes and boots. Leather can be recycled or reused, for example, you can take your old shoes to a **shoe bank**.

Saving clothes

Reusing and recycling our old clothes saves **raw materials** such as cotton and wool and saves **energy**, such as **electricity** in factories. It also cuts down on the **pollution** made by clothes factories, for example when textiles are **dyed** and **bleached**.

◄ Cotton fibres are found in the seed pods of the cotton plant. Cotton crops are grown around the world.

Factories can also reduce clothes waste. Leftover threads can be collected and reused. Unwanted scraps of material, called 'offcuts', can be sold and reused for crafts such as rug making. Scraps can also be shredded to make stuffing for mattresses and furniture.

▲ This man is using 'offcuts' to make some furniture cushions.

LOOKING AFTER YOUR CLOTHES

1. Wear an apron or an old shirt if you are doing something messy.

2. Wear old clothes for playing outside.

Wear it again

There are many ways of reusing clothes you don't want anymore. You could give them to a younger member of your family. If you and your friends no longer want some of your clothes, ask an adult if you can hold a clothes swapping party.

► *When you grow out of your clothes, pass them on to someone smaller than you so they can be reused.*

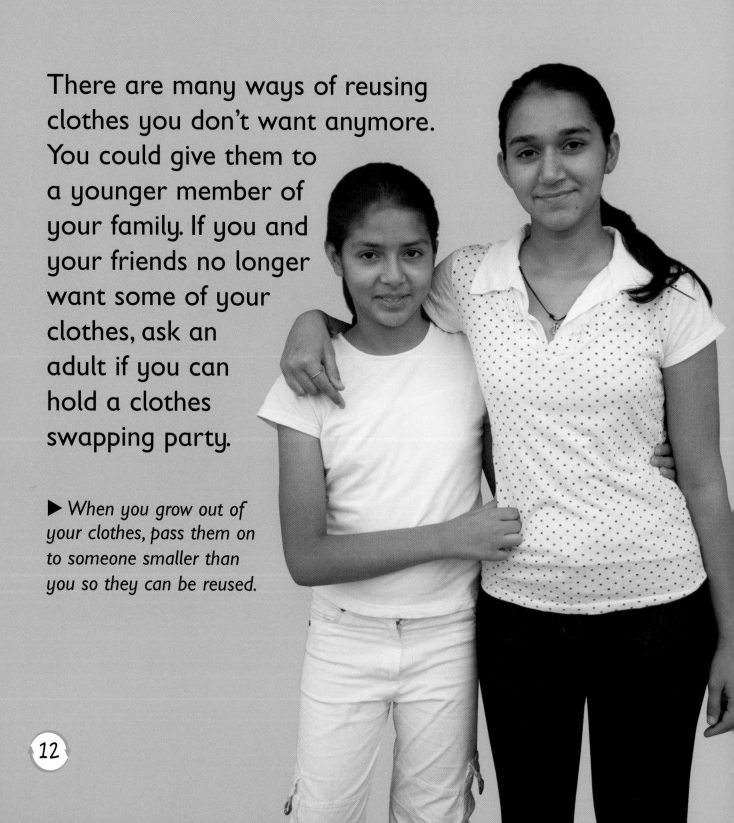

▶ Whenever you can, buy clothes from charity shops. You are reusing clothes, and giving money to help people in need.

You Can Help!

Before you give your clothes away to a charity shop, wash and fold them, and put them in a bag to keep them in a good condition.

Unwanted clothes can be given to a **charity shop**. Some may be sold at the shop. Others will be given away to homeless people or to people who have been hurt in a **natural disaster**. They may also be given to poor people in Africa and Asia.

Recycling clothes

When we recycle our clothes, we turn waste clothes into new things. Threads are taken from old clothes and used to make new clothes, or other useful things. Woolly clothes, for example, can be sorted into different colours so the wool may be used again.

◄ *The wool from this knitted jumper could be* **unravelled** *to make a blanket or a scarf.*

Did You Know?

New clothes can be made from recycled plastic bottles. It takes 25 big drinks bottles to make one large fleece.

In the UK, only a quarter of the space in clothes banks is used. Ask an adult to help you find your nearest clothes bank. Clothes from the bank are taken away to be sorted. Damaged clothes may be sold to factories, cut up and made into cleaning cloths.

◀ Keep a basket at home for any clothes that cannot be reused. When it is full, take it to the clothes bank.

Too many toys

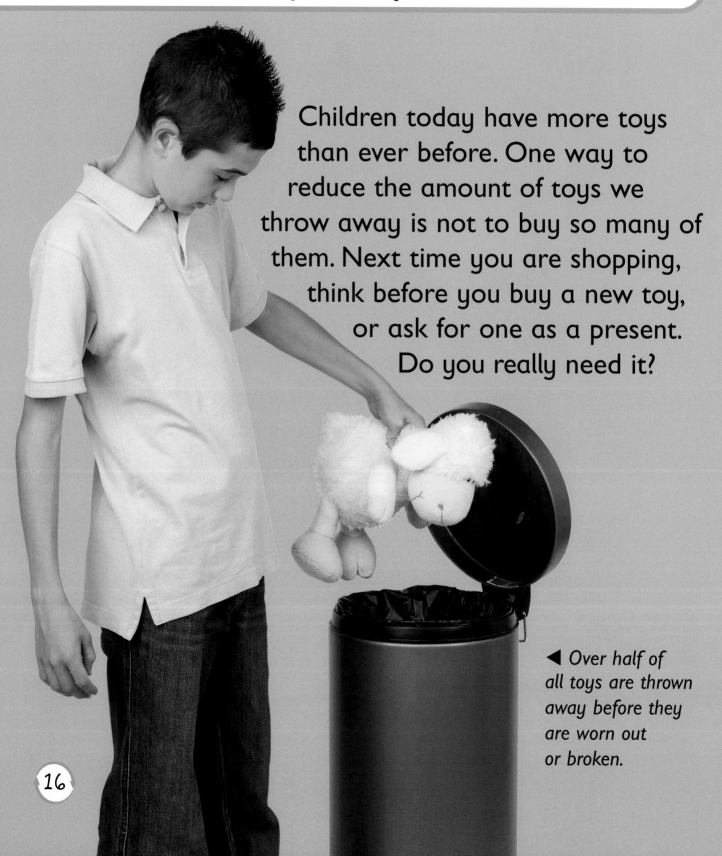

Children today have more toys than ever before. One way to reduce the amount of toys we throw away is not to buy so many of them. Next time you are shopping, think before you buy a new toy, or ask for one as a present. Do you really need it?

◄ *Over half of all toys are thrown away before they are worn out or broken.*

There are lots of reasons to buy fewer new toys. Some toys travel a long way from the country where they were made to the place where they are sold. This uses lots of **fuel** and energy, and causes pollution which is bad for the **environment**.

▲ Look on the packaging. It tells you where it was made.

Did You Know?

Eight out of ten toys sold around the world are made in China. A doll made in China travels 7,199 miles to be sold in a shop in New York, United States of America (USA), or 5,059 miles to reach London in the UK.

Toy materials

Toys are made from many different materials. Some are made from natural materials like wood and some are made from man-made materials, such as fur fabric or plastic. When plastic toys are dumped in landfill sites, they take 500 years to **decay.**

▲ *Most cuddly toys are made with fur fabric. This is a man-made material.*

Many electronic toys and games are made with materials that can be recycled. Old games consoles and games do not have to be thrown away. Shops will often buy old games from you to resell and may also be able to tell you where they can be recycled.

▲ If you throw an electronic toy away, take out the batteries first so they can be recycled.

BATTERIES

1. Don't throw batteries away in your bin. Batteries have **chemicals** inside them that can cause pollution.

2. You can take old batteries to special local recycling points.

3. Use **rechargeable** batteries for your electronic toys and games, so you don't keep buying and recycling batteries.

4. Ask an adult to connect your electronic game to mains electricity so you don't need batteries at all.

Toy packaging

Many new toys are wrapped in strong, bulky packaging. Some packaging stops toys from getting broken when they travel from the factory to the shop. But a lot of it is unnecessary. Making packaging in factories also uses energy.

◀ This shop has lots of toys wrapped in packaging. Most of this packaging will end up as waste.

When toy packaging is thrown away, it usually ends up in landfill sites. Toy packaging is made from different materials, for example paper, cardboard and plastic. Most of these materials can be easily recycled, but they must be separated and sorted first.

◀ After a birthday party, sort out your toy packaging carefully. Save all the parts that can be recycled.

Pass it on

Reusing toys may mean people buy fewer new ones. Look after your toys, so they don't break. That way you will be able to pass them on to someone else.

▲ Clean your toys before you give them away and ask an adult to check they are working properly.

You could **donate** your old toys to a charity shop. By giving your toys to charity, someone else in another country will be able to reuse your unwanted toys.

▲ This little girl in Africa is playing with a toy donated by charity.

HAVE A CLEAR-OUT

1. You could give your toys to friends or to members of your family.

2. Ask a parent to donate some of your toys to the children's ward at your nearest hospital.

3. You could organise a toy stall at your school fête to sell unwanted toys.

Sell, swap or borrow

You can make sure your old toys are reused by selling them. This is a great way to earn some extra pocket money. If you no longer want a toy, ask an adult if you can swap toys with a friend. That way, both your toys are reused.

◀ *Always check that the toys you want to sell are in a good condition.*

MAKE SOME MONEY

1. Try organising a sale outside your house.

2. You could choose to sell your old toys at a boot fair.

3. Ask an adult to help you sell your toys **online**.

Ask your teacher if you can start up a toy swapping club at school. Another way to reuse toys is to join a toy library. Toy libraries are just like ordinary libraries, except you borrow toys, not books. You can also stay in the library to play with the toys, too.

▼ Toy libraries are good places to borrow all sorts of toys.

Toy recycling

If you break a toy, ask an adult to mend it for you. If it can't be mended, perhaps it can be recycled. Look at the material your old toy is made of. Wood and some kinds of metal can be recycled, so take toys made from these materials to a recycling centre.

▼ *This woman is sorting through books and toys to see which ones can be recycled.*

You can make new toys from recycled materials, for example, you can cut up an old fleece or furry jacket to make a soft toy. You can also make accessories for your toys out of toy packaging, for example you could use a box to make a home for one of your favourite toys.

▲ These cars are made from aluminium cans.

TRY SOME JUNK-MODELLING

1. Make a rag doll from old scraps of material.

2. Use an odd sock to make a sock puppet.

3. Make a pom-pom from some old wool.

4. Cut up some plastic bags to make a kite, using thin wooden rods to make the frame.

Make a multi-coloured coat

Second-hand clothes can be made into something new, for example it is easy to make a coat from an old shirt. You could sew some more material onto the bottom of the shirt to make it longer. Ask an adult to help you do this.

YOU WILL NEED:

1. a large old shirt,
2. some scraps of brightly-coloured material,
3. paper, pencils and felt-tip pens,
4. safety scissors,
5. pins,
6. PVA glue.

1. Start making your coat by cutting the collar and cuffs off the shirt.

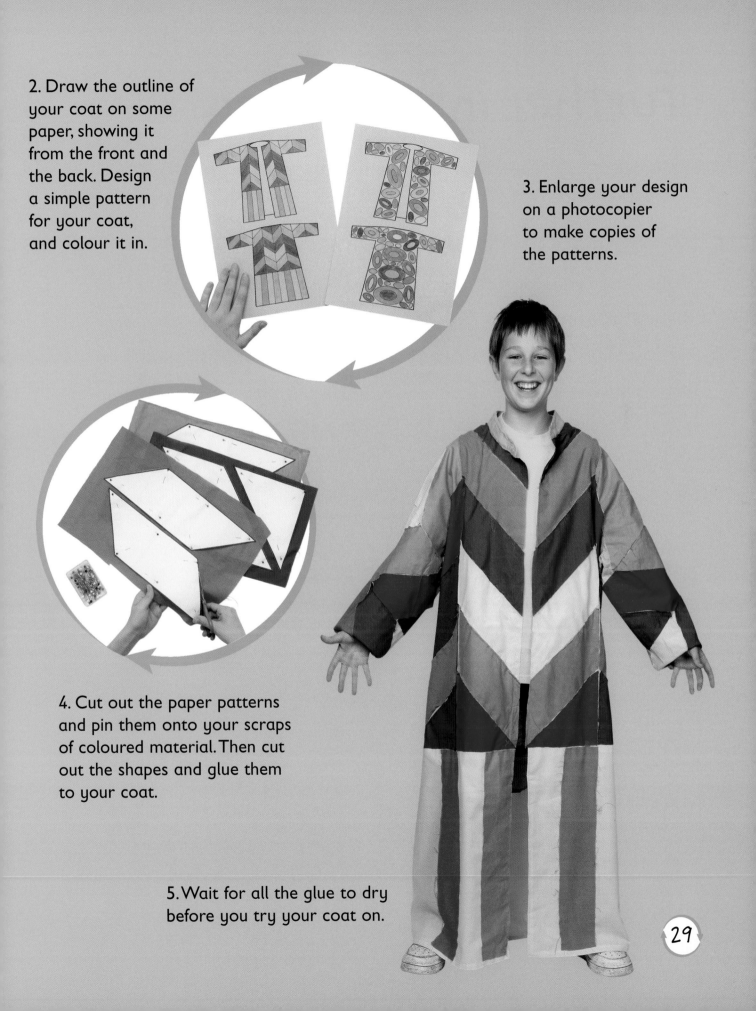

2. Draw the outline of your coat on some paper, showing it from the front and the back. Design a simple pattern for your coat, and colour it in.

3. Enlarge your design on a photocopier to make copies of the patterns.

4. Cut out the paper patterns and pin them onto your scraps of coloured material. Then cut out the shapes and glue them to your coat.

5. Wait for all the glue to dry before you try your coat on.

Further information

Topic map

SCIENCE

Dig two holes in the garden. Bury a wooden lolly stick in one hole, and a plastic straw in the other. After three months, dig up the stick and the straw. You'll see that wood rots quickly but plastic doesn't.

MATHS

Hold a toy sale and invite your friends. Put price labels on all the toys, and make sure you give the right change for them. Add up all the money you make at the sale.

ENGLISH

Write a story about a toy that is saved from the rubbish bin and sold at a charity shop. Give your story a sad beginning, and a happy ending!

GEOGRAPHY

Find out about a recent natural disaster, such as a flood or an earthquake. Where did it happen? Did any charities send food and clothes to help the people who were affected?

ART

Customise a pair of jeans and a T-shirt from a charity shop. Ask an adult to help you use fabric paint, and glue on patches to create your own style.

PSHE

Start a toy swapping club at school. Encourage your friends to share and reuse toys, rather than buying new ones.

Further reading

Environment Action: Recycle by Kay Barnham (Wayland, 2008)
Improving Our Environment: Waste and Recycling by Carol Inskipp (Wayland, 2005)
Re-using and Recycling by Ruth Thomson (Franklin Watts, 2006)

Websites

www.e4s.org.uk/textilesonline/world.htm
Games and facts about textiles and recycling.

www.olliesworld.com
Children's website with information on recycling in USA, Australia and UK. Includes a section on textiles.

www.recyclenow.com
Covers recycling in all areas of life, eg home and school.

www.natll.org.uk
Visit the National Association of Toy and Leisure Libraries' website to find your local toy library.

Glossary

bleached to have colour removed

charity shop a shop that sells second-hand things to raise money

chemical a substance, or mixture of different substances

clothes bank a place where you can leave clothes to be recycled

condition the state of something

decay to rot or go bad

donate giving items away to charity

dyed adding colour to a material

electricity a kind of energy used to make things work

energy the power to work

environment the world around us, and all living things

fuel something that is burnt to make power

furnace a hot oven in which things are burned

incinerator a very hot furnace in which rubbish is burned

landfill site a big rubbish dump, where waste is buried underground

man-made something made, that is not found naturally

material cloth or fabric

natural something found naturally

natural disaster an event of nature e.g. a hurricane

online using the Internet

pollution dirt in the air, water or earth

population the people who live in a place

raw material natural materials

rechargeable a source of energy that can be reused

recycle when something is made into a new product

recycling bank a place where you can leave items to be recycled

recycling centre a centre where waste materials are recycled

reduce to cut back or make smaller

reuse when something is used again

shoe bank where shoes can be dropped off to be reused or recycled

textile a type of cloth made from threads or woven fabric

unravelled undo something that is knitted together

31

Index

Numbers in **bold** refer to a photograph.

b
battery 19, **19**
boot 9
boot fair 25
c
charity shop 13, **13**, 23, 30
clothes 5, 6, **6**, 7, 8-9, **8**, 10-11, 12,
  13, 14-15, 28, 30
clothes bank 7, **7**, 15
cotton 8, 10, **10**
e
electronic toy 19, **19**
f
factory 10, 11, 15, 20
family 12, 23
fleece 15, 27
friend 12, 23, 24, 30
g
games console 19
h
homeless 13
l
landfill site 4, 18, 21
m
man-made 8, 18
material 8-9, 10, 11, 18-19, 21, 26, 27,
  28, 29
n
natural 8, 9, 18
natural disaster 13, 30

o
online 25, 30
p
plastic 15, 18, 21, 30
plastic bag 27
r
raw material 10
rechargeable battery 19
recycle 5, 6-7, 8, 9, 10, 14-15, 19, 21,
  26-27, 30
recycling bank 7
recycling centre 7, 8, **8**, 26
reduce 5, 6-7, 11, 16
reuse 5, 6-7, 9, 10, 11, 12, 13, 15, 22, 23, 24,
  25, 30
s
school 23, 25, 30
second-hand 7, 28
shoe 9, **9**
shoe bank 9
shop 17, 19, 20, 23
shopping 16
t
textile 8, 10, 30
thread 8, 11, 14
toy 4, 5, 6, 7, 16-17, **16**, 18-19, **18**, 20,
  22, **22**, 23, 24, 25, 26-27, **26**, 30
toy library 25, **25**
toy packaging 17, 20-21, **20-21**, 27
toy stall 23, 24